The Life and Death of the Universe: The History of the Big Bang and the Ultimate Fate of the Universe

By Sarah Malloy & Charles River Editors

A Hubble telescope picture of various galaxies in the universe

About Charles River Editors

Charles River Editors is a boutique digital publishing company, specializing in bringing history back to life with educational and engaging books on a wide range of topics. Keep up to date with our new and free offerings with this 5 second sign up on our weekly mailing list, and visit Our Kindle Author Page to see other recently published Kindle titles.

We make these books for you and always want to know our readers' opinions, so we encourage you to leave reviews and look forward to publishing new and exciting titles each week.

Introduction

Artist's depiction of a satellite gathering information about the Big Bang

The Universe

"You, King Gelon, are aware the Universe is the name given by most astronomers to the sphere the center of which is the center of the Earth, while its radius is equal to the straight line between the center of the Sun and the center of the Earth. This is the common account as you have heard from astronomers. But Aristarchus has brought out a book consisting of certain hypotheses, wherein it appears, as a consequence of the assumptions made, that the Universe is many times greater than the Universe just mentioned. His hypotheses are that the fixed stars and the Sun remain unmoved, that the Earth revolves about the Sun on the circumference of a circle, the Sun lying in the middle of the orbit, and that the sphere of fixed stars, situated about the same center as the Sun, is so great that the circle in which he supposes the Earth to revolve bears such a proportion to the distance of the fixed stars as the center of the sphere bears to its surface." - Archimedes

Many of us on Earth live in a home with a street address, city, and zip code. But we have another address, which astronomers and astrophysicists refer to as our "cosmic address": Earth, Solar System, Milky Way, Local Group, Virgo Supercluster, the Universe. Consider the vast scales that our cosmic address spans. Understandably, it's hard for our brains to truly imagine it. Our galaxy, The Milky Way, is populated with somewhere between 100-400 billion stars, one of which is our Sun, an average, run-of-the-mill star. The nearest star to us is Proxima Centauri at a distance of 4.2 light years or almost 4×10^{13} kilometers. A light year is a measure of distance that it takes a photon (particle of light) to travel in a year, assuming it is traveling in the vacuum of space. So, it takes light 4.2 years to travel from Earth to Proxima Centauri and it would take a space shuttle about 160,000 years to get there, assuming our shuttle was traveling at a little over 4,000 miles per her. Considering that traveling to our nearest stellar neighbor is well outside of our technological capabilities and it would take more than 1600 times a human lifespan to get there, space travel either to or from other habitable planets is extremely improbable in the near future.

We live in the outer edge of one of the spirals of the Milky Way Galaxy. The image[1] below is an artist representation of the disk of our Galaxy, showing a small dot where our Solar System lies.

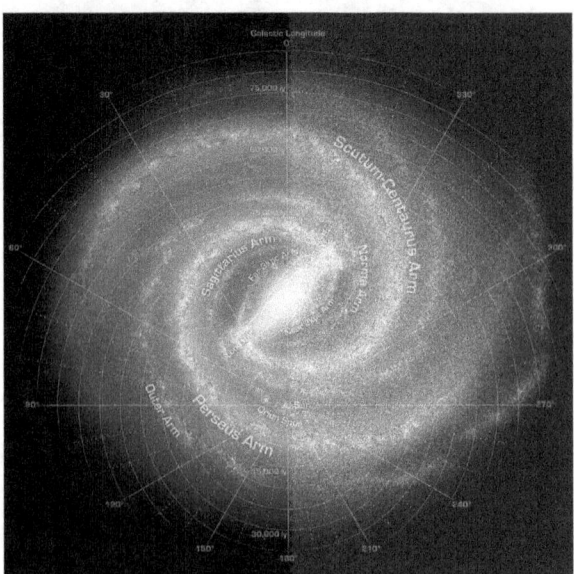

Diagram of the Milk Way, courtesy of NASA/Jet Propulsion Laboratory

The diameter of the Galaxy is about 100,000 light years across, so while it takes 4.2 years for

---

[1] Cain, Fraser. The Universe Today. "Where is Earth in the Milky Way?" May 31, 2010
<http://www.universetoday.com/65601/where-is-earth-in-the-milky-way/>

light to travel to our nearest neighbor, it takes light 100 millennia to travel to the other side of the Galaxy. The universe is estimated to contain at least 200 billion galaxies, according to observations possible with today's technology[2]. When you look at any object at a distance, you are looking into the past. You are seeing the photons as they were when they scattered, radiated, or reflected off of that object. Let's consider the star Proxima Centauri. The photons that leave the surface of Proxima Centauri have a 4.2 yr travel time until they reach Earth. In other words, when we look at Proxima Centauri, the light that is hitting our telescope is 4.2 years old. Another example is our nearest major galactic neighbor, the Andromeda Galaxy, which is 2.5 million light years from us. That means that when we observe Andromeda, we are looking back in time 2.5 million years ago, about the time when the earliest known ancestors of ours are suspected to live on Earth. The furthest[3] galaxy that we have ever observed is 13.4 billion light years away, and since the age of the universe is estimated to be 13.7 billion years old, that means that we are receiving photons from a galaxy that formed very soon after the big bang.

Experiments done in astronomy are quite different from other sciences. Unlike biology, chemistry, and physics, astronomers cannot hold their specimens in a petri dish and put them under a microscope. We rely solely on electromagnetic radiation to travel through space, into our telescopes, and interact with our detectors. While there have been impressive advances in our understanding of astronomy and the physical world, we are still quite limited by the sensitivity of our instruments, the size of our detectors, and our inability to probe the far reaches of the cosmos. As such, scientists can only attempt to answer questions that are within grasp of our current technology. Wondering what happened before our universe existed is a matter best left for dreaming, since making such observations are not (yet) possible.

The Life and Death of the Universe: The History of the Big Bang and the Ultimate Fate of the Universe examines the fascinating history of deep space, and what the future holds. Along with pictures of important people, places, and events, you will learn about the universe like never before.

---

[2] The Hubble Space Telescope has mapped small regions of the sky, and scientists use software to count sources of light that appear to be galaxies, and then extrapolate over the entire sky to estimate the total number of galaxies.
[3] Hubble Space Telescope team reported this observation in March 2016 and more information can be found at http://hubblesite.org/newscenter/archive/releases/cosmology/distant%20galaxies/2016/07/full/

The Life and Death of the Universe: The History of the Big Bang and the Ultimate Fate of the Universe
About Charles River Editors
Introduction
   The Electromagnetic Spectrum
   Olbers' Paradox
   Discovering Our Place in the Universe
   The Great Debate
   The Big Bang Theory
   Doppler Shift
   Hubble's Law
   Penzias and Wilson
   The Cosmic Microwave Background
   The Cosmological Principle
   Estimating the Age of the Universe
   Abundance of Light Elements
   Inflation Theory
   Large Scale Structure
   The Horizon Problem
   The Flatness Problem
   Fate of the Universe
   Alternate Theories
   Dark Matter and Dark Energy
   Future Experiments
   Conclusion
   Online Resources
   Bibliography
Free Books by Charles River Editors
Discounted Books by Charles River Editors

The Electromagnetic Spectrum

Light is the primary medium that we use to perform experiments in astronomy outside of our solar system. We define light as electromagnetic radiation, and visible light is of wavelengths between 400 and 700 nanometers; it is this range in which human eyes are sensitive. For example, when we shine a light on an object in a dark room, it becomes visible to us because light is being reflected off of the object and into our eyes. Of course, the kind of radiation humans see is only a tiny sliver of the electromagnetic spectrum, but luckily we ca. The figure[4] below shows the full electromagnetic spectrum from gamma rays to radio waves, with the representation of wavelength, frequency, and energy below it. For reference, visible light is shown as the rainbow, spanning the colors red, orange, yellow, green, blue, indigo, and violet. All other parts of the spectrum are invisible to us.

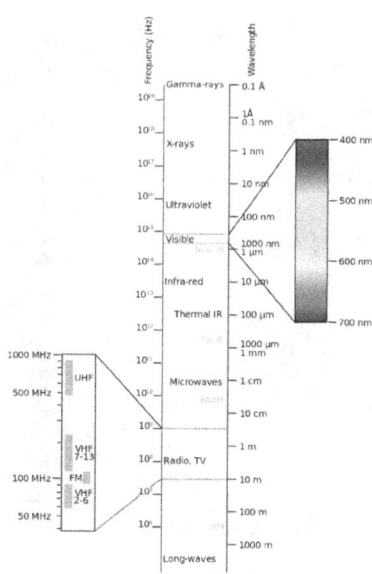

Electromagnetic spectrum, courtesy of Victor Blacus

A photon is a particle of light (yes, light is both wave-like a particle-like), and an electromagnetic wave is a packet of photons. The electromagnetic spectrum encompasses all of the different energies (or analogously, wavelengths or frequencies) radiation can have. Electromagnetic radiation can be expressed in energy (E), frequency (ν), or wavelength (λ) by the equations:

---

[4] Comparison of wavelength, frequency and energy for the electromagnetic spectrum (Credit: NASA's Imagine the Universe).

$$v = \frac{c}{\lambda} \qquad E = \frac{hc}{\lambda}$$

The two constants are the speed of light, $c = 3 \times 10^8$ m/s, and Planck's constant, $h = 6.626 \times 10^{-34}$ m²kg/s. The units astronomers use for frequency is Hertz, wavelength is in meters, and energy is in electron volts. The electromagnetic spectrum is composed of gamma rays ($\lambda \sim 1 \times 10^{-12}$ meters or 1 picometer), X-rays ($\lambda \sim 1 \times 10^{-10}$ meters), Ultra-violet ($\lambda \sim 100 \times 10^{-9}$ meters or 100 nanometers), Visible (λ is 400-700 nanometers), Infrared ($\lambda \sim 100 \times 10^{-6}$ meters or 100 microns), microwaves ($\lambda \sim$ 1 centimeter to 1 meter), and radio waves ($\lambda \sim$ 100 meters to 100 kilometers). Gamma rays have the shortest wavelength, the highest frequency, and the highest energy while radio waves have the longest wavelength, the lowest frequency, and the lowest energy.

Max Planck, the Nobel Prize Winning Physicist

Gamma rays, X-rays, and ultraviolet rays are all blocked by the upper atmosphere of Earth, so it is best to observe those types of photons from space. Visible light penetrates the atmosphere, and we can easily observe visible light from the ground. The infrared part of the spectrum is absorbed by gasses in the atmosphere and so most infrared observatories are designed to be high in the atmosphere or in space. Radio waves completely penetrate the atmosphere and all of the radio observing is done from the ground. The light we receive from a star gives us clues about the properties of the star, since the energy of the light does not change when traveling through the vacuum of space. A few examples of what we can learn from the light we receive from a star: (1) we get information about the flux or brightness of the star, (2) we can derive its distance, (3) we can use spectroscopy to determine its chemical composition, and (4) we can tell if it is moving toward or away from us by measuring shifts in its spectral lines (discussed in the section on Doppler shift).

Olbers Paradox

In 1826, astronomer Heinrich Olbers posed the question, "why is the sky dark at night if the universe is infinitely old and there are an infinite number of stars?" If it were true that there are an infinite number of stars distributed throughout the universe, then every line of sight would end on a star. Olbers broke the observable universe up into concentric shells of a sphere, so a certain number of stars reside in a shell from 1-2 light years away, 2-3 light years away, and so on. The shells that are further away from us cover more area, so they hold more stars. The Earth would receive the same amount of light from each shell, even the shells that are further away since they contain more stars. So the more shells there are going further and further into the assumed infinite universe, the more light Olbers assumed we should receive. Given this assumption, one would expect that our night sky would be at least as bright as the Sun.

The solution to this paradox is that the universe is finite, the speed of light is finite, and there are a finite number of stars. Furthermore, since the universe is expanding, light traveling to us appears to have energy levels shifted into the red part of the spectrum, a concept called redshift, which will be discussed in a later section. The solution to Olbers paradox is that (1) stars are much less dense than Olbers assumed, (2) the expansion of the universe dampens the energy of light we receive from them, (3) light travels at a finite speed, and (4) the total amount of space in the universe is finite. The reality that we know well today is that the universe overall is very dark and composed mostly of empty space.

Discovering Our Place in the Universe

Aristotle's model of the universe was based on the Earth being at the center of everything, and that was the accepted model from 322 BC to the 16th century. The geocentric model proposed by Aristotle puts the Earth at the center of the universe with the Sun, planets in our Solar System, and everything else in the universe revolving around it. The universe was thought to be finite, bounded, and static. In the mid-1500's, Copernicus proposed[5] that the Sun is at the center of our universe, which challenged many fundamental ideas in cosmology, philosophy, and religion. Copernicus posed the question "if the Earth is not at the center of the universe, why should any place be at the center?" or "why should the universe even have a center?" His sun-centered or heliocentric model caused a major shift in thought and impacted scientific theories for more than a century after his publication, this shift in thought is called the Copernican Revolution.

---

[5] Nicolaus Copernicus' heliocentric model was published in the year 1543 in "de revolutionibus orbium coelestium" translated to "On the Revolutions of the Celestial Spheres."

Nicolaus Copernicus

In 1609, Johannes Kepler published his first two laws of planetary motion, which is based on the Sun being at the center of the Solar System and orbits no longer being perfect circles. Around the same time, Galileo built a state-of-the-art telescope and observed Jupiter's moons orbiting around it. This caused another huge ripple in scientific thought because Galileo discovered that there was another heavenly body with smaller planets (moons) orbiting around it, proving that there are heavenly bodies not orbiting Earth. Galileo's support of the heliocentric model ultimately summoned him to house arrest by the catholic church because scripture disagreed with his scientific research. He was punished for having the opinion that the Sun was at the center of the universe instead of the Earth. His books and any future writing was banned, and he was summoned to house arrest for the rest of his life. He died there at age 77.

Galileo Galilei

Modern cosmology holds the idea that there is no center of the universe. A good analogy is to represent the universe as the surface of a balloon, breaking a 3-dimensional universe down into a 2-dimensional one. [6] If you draw dots on the balloon to represent galaxies, the space between the galaxies gets larger as we blow up the balloon. From any point on the balloon, all other points appear to be moving away. If you lived on one of those points, you might believe you are at the center because every point is moving away from you. Considering the balloon as a whole, there is no center on the 2-dimensional balloon universe. Space is expanding everywhere. The big bang is not an explosion that can be traced back to a point, rather it occurred everywhere in the universe at once, like the surface of the balloon. The observable universe is defined by the observer's location, so we are at the center of our observable universe.

The Great Debate

The great debate was a meeting of the National Academy of Sciences in Washington on April 26, 1920. The debate was on the scale and makeup of the universe and more specifically, whether the "spiral nebulae" were other galaxies at great distances from us or if they were clumps of gas within the Milky Way galaxy. Before the 1920's, the ideas on the evolution and makeup of the universe were quite different, and the great debate settled some very important questions about our galaxy and the structure of the universe at large. There were two prominent astronomers leading the debate, Harlow Shapley and Heber Curtis.

---

[6] Stephen Hawking, "A Brief History of Time", 1988.

Shapley held the belief that the universe was made up of our galaxy, and that everything that we observe is within the bounds of the Milky Way. He believed that the spiral nebulae, which he thought were clouds of gas that were possibly forming planetary systems, were part of our galaxy. Shapley tried to estimate the diameter of our galaxy based on the previously known distance to a globular cluster. A globular cluster is a large collection of stars that are very dense in the center with a decreasing density as distance from the center of the cluster increases. He believed that globular clusters made up the halo of our galaxy, which he was right about, but assumed, wrongly, that all globular clusters were the same size. Using these assumptions, he concluded that our galaxy must be 300,000 light years in diameter, and that the Sun is 50,000 light years from the center.

On the other side of the debate, Curtis estimated the diameter of our galaxy to be 30,000 light years, much smaller than Shapley's estimate, and that the Sun was at the center. He believed that the spiral nebulae were other galaxies outside of our galaxy, calling them "island universes". He based his opinion on the similarity of appearance (disks with obscuring material) between spiral nebulae and our galaxy, and the similarity of the spectra (the chemical fingerprint) of the spiral nebulae and our galaxy. Curtis also thought that globular clusters made up the halo of the Milky Way, but he disagreed with Shapley's estimate of the distance to them.

The great debate was not settled until Edwin Hubble used a better telescope, the 100-inch Mt. Wilson Observatory, to use better techniques to measure the distance to stars in the Andromeda galaxy, our nearest major galactic neighbor. Hubble's story will be explained in detail in the following section on Hubble's Law. To conclude the debate, both Shapley and Curtis were correct about some things, but they both were missing pieces to their stories. Curtis was correct that the "island nebulae" were other galaxies far outside of the Milky Way. Shapley was correct that the Sun is not at the center of our galaxy, and his estimates of the size of our galaxy and location of the Sun within our galaxy were much closer to today's commonly accepted values.

With a better understanding of our place in the universe, how astronomers use light to understand physical properties of distant objects, and that the universe is composed of a plethora of empty space, with galaxies hosting stars and planets just like our own, let's move on to the most plausible theory for how this universe came to be.

The Big Bang Theory

The prevailing cosmological model for the evolution of the universe is the Big Bang theory, which describes:
1. The evolution of the universe from the earliest moments
2. How the expansion and cooling of particles and photons could have led to the present universe of stars and galaxies
3. The formation of light elements (e.g. hydrogen, helium, lithium)

4. The large scale structure of the universe

The big bang theory was first proposed by the physicist and priest, George Lemaitre, in the 1920's. He was the first to suggest the idea that the universe is expanding, and derive Hubble's law, attempting the first observation of what is now known as the Hubble constant (discussed in a later section). Lemaitre was the first to hold the theory that the universe originated from a single point, which he called the "primeval atom". Much later, his theory became known as the "Big Bang theory" when the astronomer Fred Hoyle coined the term in a BBC radio broadcast in 1949.

The big bang theory, as it stands today, has had many contributors and has come a long way since Lemaitre. The theory starts at 10^{-43} seconds after the big bang. At this time, three out of the four forces in the universe are unified, and gravity has separated out from the super-force. The four forces that control the physics of how all things interact are gravity, the strong force, the weak force, and the electromagnetic force. Gravity is the familiar force that we interact with every day, it is what keeps us planted on the Earth. Gravity is an attractive force that is weaker than the other forces, has a very long range, and acts between any two masses in the universe. The weak force has a short range and as its name suggests, is weak in strength. The range for the weak force is about 0.1% of the diameter of a proton. The weak force changes one type of particle into the other, for example, there are interactions in the Sun that change protons into neutrons because of the weak force. The strong force has the most strength and a short range, it acts over ranges that are about the diameter of a nucleus. The strong force is responsible for holding the nucleus of atoms together. Finally, the electromagnetic force controls the physics of charges and magnetic fields, and has a similar inverse square law to gravity. The electromagnetic force is very long range, like gravity, and it is the force that holds atoms and molecules together. According to the big bang theory, the grand unification epoch is the time in the universe when the strong, weak, and electromagnetic forces were all one force, and gravity was separated out. There is not much that is known about the universe before this because the physics that we know about breaks down.

In the next small fractions of a second, at 10^{-35} seconds and a temperature of 10^{27} K, the force of gravity becomes prominent and the strong and electro-weak forces separate, triggering an abrupt expansion that we call inflation. This short time period is appropriately called the inflationary epoch, where the universe grows by a factor of about 10^{43} in size in fractions of second. The theory of inflation has explained why the universe appears to be smooth and flat, and how widely separated parts of our universe appear to be in equilibrium. The next moments after the big bang, when rapid expansion slows down, mass appears spontaneously from energy and this mass-energy conversion causes the formation of new particles. Once the universe cools to 10^{12} K, the conversions between matter and antimatter[7] can no longer sustain and the amount

---

[7] Electrons are building blocks of matter and antielectrons, which have properties exactly like electrons except with a positive charge, are the building blocks of antimatter. All particles have antiparticle counterparts.

of matter in the universe is set. At this point, about 10^{-6} seconds after the big bang, the universe contains matter that we would recognize, namely protons and neutrons. In the next few minutes, protons and neutrons begin to fuse together, however the universe is still too hot for them to remain bonded, causing them to repeatedly fuse and then break apart. At three minutes after the big bang, the density decreases and the temperature decreases to 10^9 K, slowing fusion of protons and neutrons to a halt, leaving the first chemical imprint of 75% hydrogen and 25% helium with trace amounts of heavier elements. This time period is called the epoch of nucleosynthesis. From about 5 minutes to 380,000 years after the big bang, the universe consists of a hot, cosmic soup of hydrogen and helium nuclei and free electrons. This cosmic soup is too hot for nuclei to hold on to electrons, so photons randomly bounce off of the free electrons and quickly break apart any atoms that form. When the universe cools down to 3000 K, hydrogen and helium nuclei can hold on to electrons forming stable atoms for the first time. This time period is called the epoch of recombination. Once neutral atoms are formed and electrons are bound to nuclei, photons are allowed to travel freely and the imprint of the photons last scattering event is set. The epoch of recombination is the earliest time that we can use photons to directly observe the universe.

The figure[8] below illustrates an overview of the major events described in the evolution of the universe according to the big bang: Very early in the universe (left side the illustration), quantum fluctuations in space-time lead to extremely rapid growth of space in a short span of time.

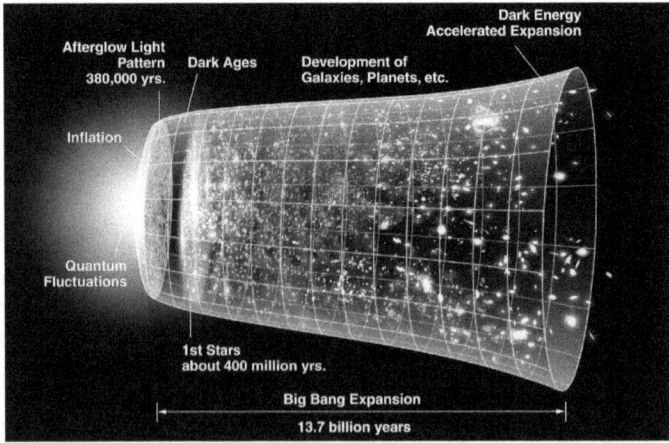

Representation of the Big Bang Theory

---

[8] Artist representation of the history of the universe according to the Big Bang model. Image Credit: WMAP Science Team/NASA

This inflationary period led to cooling allowing photons to freely escape at about 380,000 years after the big bang, producing an observable afterglow light pattern (a cosmic microwave background). From 380,000 years to 400 million years, when the first stars were formed, there were no new sources of light or interactions that give us any clues as to what was going on. Therefore, this time period is called the Dark Ages, and it is analogous to observing an ultrasound image of a fetus and a picture of a teenager, and then trying to assume without any more information what happened in between. There isn't much that is known about the Dark Ages, but there are projects dedicated to finding some ways to probe this era. After 400,000 million years to 1 billion years after the big bang, the first stars and galaxies formed and evolved into the universe we know today. The current state of the universe is 13.7 billion years after the big bang. The big bang model does remarkably well at explaining observations of Hubble's Law, the cosmic microwave background, the abundance of light elements, and the large scale structure of the universe.

Doppler Shift

Before diving into Hubble's Law, there is an important concept to understand: Doppler shift, named after the scientist, Christian Doppler, who first proposed it in 1842. The Doppler shift is a change in frequency (or wavelength) in a wave as it moves toward or away from the observer.

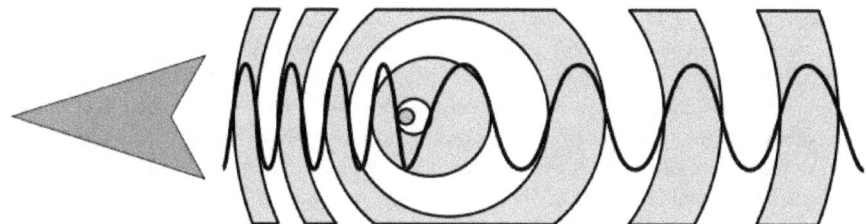

The Doppler Effect, Illustrated by a Source Moving from Right to Left. Courtesy of Tkarcher

Frequency, ν, and wavelength, λ, are inversely related by the simple equation $\lambda = \frac{c}{\nu}$, where c is the speed of light. So, light with a longer wavelength will have a lower frequency, and high frequency waves will have a short wavelength. You experience Doppler shift every day, for example when an ambulance drives by you, its pitch sounds different when it's coming towards you versus driving away from you. The image[9] below shows this concept. The person on the

---

[9] Broome, Madeleine. "Redshift: The Universe's Doppler Effect", December 9, 2015, Princeton Journal of Science and Technology

left observes the ambulance moving away from him and the person on the right observes the ambulance coming towards him, and so the relative motion of the ambulance is different for the two observers. The sound waves heard by the observers are shown as circles around the ambulance. The person on the left hears sound waves given off by the ambulance, but the waves are stretched out as a result of its relative motion. The person on the right hears the sound waves given off by the ambulance, but the waves are scrunched up as a result of its relative motion. The result is the person on the left hears a lower frequency (lower pitch) and the person on the right hears a higher frequency (higher pitch). The Doppler effect is dependent upon relative motions, meaning that it depends on the observer. The sound waves given off by the ambulance in its own frame of reference would not be shifted at all, and would appear as perfect circles in the image above. If the observers were moving, the Doppler shift would depend on the velocity of the observers relative to the velocity of the ambulance. There are many other examples in everyday life of the Doppler effect, and it occurs with anything that has a wave. A duck swimming in a pond would have scrunched up waves in front of it and stretched out waves behind it. A sonic boom is the Doppler effect in action as well, it is a shock wave given off by a supersonic aircraft catching up with its own sound waves. The Doppler radar is a device used by police officers to measure the speed of cars driving by. There are many other examples and application of the Doppler effect.

Just like sound, electromagnetic waves are changed based on the relative motion between the source and the observer. Electromagnetic waves are composed of packets of photons, the carrier particle for light. As discussed earlier, the electromagnetic spectrum is composed of all of the possible wavelengths (or frequencies), including radio waves, infrared, visible, ultraviolet, x-ray, and gamma rays. We are most familiar with the visible part of the electromagnetic spectrum because that is what our eyes are sensitive to, the rainbow of colors from blue to red. If we observe a star or galaxy moving toward us, the light waves from that object are scrunched and the frequency of the wave appears higher, just as the example above with sound. The term we use to describe this is blueshifted because the light we see will look bluer. Redshift is exactly the opposite, the light waves from a source moving away from us appear to be stretched out with a lower frequency, and its light will shift towards the red.

In practice, astronomers use the Doppler effect to infer the distance to an object by measuring the shift in its spectral lines. The chemical composition of a star or galaxy produces a "chemical fingerprint", because atoms and molecules are known to absorb photons at very specific

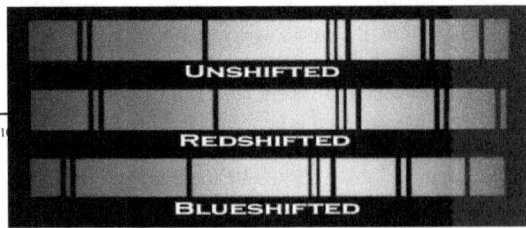

frequencies, we see that as a black line in its spectrum, as shown in the image[10] below. The spectrum shows the spectrum when the source is not um when the source is moving away from the hen the source is moving towards the observer

labeled "unshifted" is a simplified representation of an absorption line pattern of any star, galaxy, or gas cloud. We observe that pattern as either redshifted or blueshifted if we are moving away or toward each other, respectively. Once the amount of shift in frequency or wavelength is known, we can easily back out the relative velocity. The equation for redshift is given by

$$z = \frac{\lambda_o - \lambda}{\lambda}$$

where z is the redshift, λ_o is the observed wavelength, and λ is the rest wavelength.

Most of what astronomers observe is in the form of redshift, because as it turns out, most objects are moving away from one another. There are cases of blueshift, especially in local environments. When we observe two stars orbiting each other, known as a binary star system, the stars move towards us in some parts of their orbit and away from us in other parts of their orbit.

The Binary Star System Sirius, by Chris Laurel

Therefore, binary star systems have spectral lines that vary with time dependent on their orbital properties. Another example is a galaxy that is close enough to us to be affected by our mutual gravity, and therefore may be moving towards us. Redshift is much more common in the universe as a whole and there are three main types. Type I redshift is produced by the relative motion of objects, just as we have been discussing in the examples above. For example, if a star is moving away from us. Type II redshift is called the cosmological redshift, which is caused by the expansion of space itself. For example, two objects can be stationary relative to each other, but since space is expanding, it is causing them to appear to be moving away from each other just like the example with the balloon discussed earlier. Type III is gravitational redshift, caused by the gravity of a massive object shifting the frequency of the light waves emitted from a source within its gravitational field, and it is a result of gravitational time dilation derived in Einstein's theory of general relativity.

Hubble's Law

In 1929, astronomer Edwin Hubble published a paper showing that nearly all galaxies appeared to be moving away from us and from one another. His observations of distance and redshift led him to prove a very important relationship, originally derived by George Lemaitre: $v = H_o \times d$.

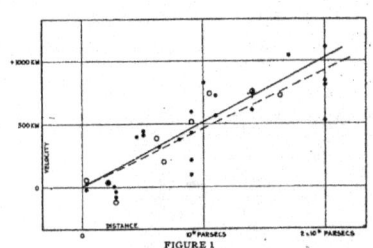

The radial velocity, v, is calculated from the amount of shift observed in the spectral lines of the galaxy, and d is the distance to the galaxy calculated from observing a predictable class of variable stars called Cepheid Variables. Hubble showed not only that the universe was expanding, but he figured out the rate of expansion, H_o, now called the Hubble constant. The image to the left shows a figure from his 1929 paper[11] where he published these results. The two lines and two dot types represent different solutions for alternate ways of measuring the distances. The trend is clear in Hubble's results, although the uncertainties in the measurements are high, he published the first decent observations measuring the expansion rate of the universe. Hubble is given all of the credit for discovering the expansion of the universe, however, George Lemaitre published a paper in 1927, two years earlier, deriving Hubble's Law. Lemaitre had poor data and he could not prove that there was a linear relationship between distance and velocity, so his work was mostly theoretical. Hubble was the first to have data that were good enough to prove the linear relationship in Hubble's Law.

Measuring the distance to astronomical objects well has only been possible in the last century. The method used to determine the distance depends on how close the object is to us, appropriately called the distance ladder. The first rung of the distance ladder is radar ranging, which is appropriate for objects out to about the distance of Saturn. Radar ranging works by reflecting microwaves off of an object (e.g. asteroid, moon, planet) and timing how long it takes for it to return. The second rung of the distance ladder is called parallax, which works for nearby stars out to about 100 light

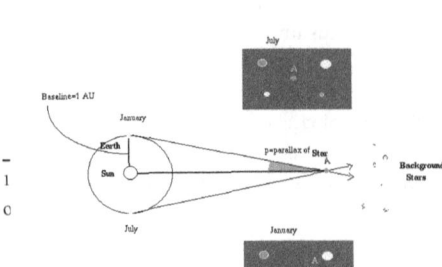

al Velocity among Extra-Galactic Nebulae", Proceedings f America, Volume 15, Issue 3, pp. 168-173

years. This method relies on the geometry of the Earth's orbit around the Sun and the resulting apparent shift in the position of the star, as shown in the figure[12] to the left. As the Earth moves in orbit from July to January, the position of Star A relative to the background (more distant) stars changes, and that shift in position can be measured in the images (shown as boxes in the figure). The background stars don't shift position in the image because they are all much further away and their shift is so small that the parallax angle is negligible. Simple geometry allows astronomers to back out the distance. The next rung of the ladder is called main-sequence fitting, and it can be used for star clusters out to 100,000 light years. A cluster of stars contains hundreds to thousands of stars at all different masses. As a whole, the collection of stars that are burning hydrogen in their core (a stage of stellar evolution called the main sequence) have a predictable luminosity. Once the luminosities of the stars in the cluster are known, we can back out the distance easily as it follows an inverse square law given by the equation,

$$B = \frac{L}{4\pi d^2}$$

where B is the apparent brightness, L is the luminosity and d is the distance. The next rung of the ladder uses a type of variable star called Cepheid variables, and this works for nearby galaxies out to 10,000,000 light years. Cepheid variables have a predictable period-luminosity relationship, pulsating with regular periods that depend intrinsically on the mass, temperature, and luminosity of the star. Stars with longer periods are brighter than stars with shorter periods. Astronomers can measure the Cepheid variable star over a few cycles to accurately measure its period, determine its luminosity from the well-known period-luminosity relationship, and then calculate the distance from the luminosity using the equation above. There are a few other methods for determining distance, one of which is using Hubble's Law by measuring the redshift of a distant galaxy. Hubble's Law works for the most distant objects in the universe.

Over the years, astronomers debated about the value of H_o. The expansion rate that Hubble himself deduced was about 500 km/s/Mpc (Mpc is mega parsec, a unit of distance equal to about 3.26 light years), and many years later, recent research[13] using the Hubble Space Telescope has settled on an expansion rate of 65 ± 10 km/s/Mpc. Hubble's law tells us that galaxies appear to be rushing away from us with a velocity that is proportional to their distance from us. In other words, galaxies that are further away (and also further back in time) appear to be moving away from us faster.

Hubble was witnessing space itself expanding, not the galaxies themselves moving with the observed radial velocity. A popular example of this concept is to think about raisin bread, where space is the bread and galaxies are represented by the raisins. Pretend that you are an ant sitting on one of the raisins. As the bread expands, all of the other raisins would appear to be moving

---

[12] Image Credit: NASA
[13] More information on the history of the Hubble Constant can be found at
https://www.cfa.harvard.edu/~dfabricant/huchra/hubble/

away from you, even though they aren't moving themselves in the bread at all. None of the raisins would move towards you. Suppose the bread doubles in size in 2 seconds, the distance between your raisin and a raisin that is initially 2 cm away ends up 4 cm away, and a raisin that is 6 cm away ends up 12 cm away. The relative velocities of those raisins are given by $v = \frac{d}{t}$, where d is the distance the raisin moved and t is the amount of time it took to move that distance. The raisin that was 2 cm away had a relative velocity of 1 cm/s and the raisin that was 6 cm away had a relative velocity of 3 cm/s. The raisin that was further away from you moved away faster, which is analogous to what Hubble discovered.

Penzias and Wilson

Two astronomers, Arno Penzias and Robert Wilson, working at Bell Labs were using a new 20-ft radio telescope to map out signals of the space between galaxies, which is a very small signal. Therefore, they had to do their best to reduce all noise or interference in the system, namely radar, radio broadcasting, and the system itself. But they could not seem to get rid of a bothersome background "noise" at a wavelength of 7.35 cm. After much trial and error, including spending hours cleaning pigeon droppings off of the antenna, Penzias and Wilson began to consider that the signal may not be noise. At about the same time, a theoretical physicist named Robert Dicke of Princeton University postulated that there should be low-level radiation leftover from the big bang, observable in every direction. Penzias and Wilson finally realized that they were observing the radiation that Dicke's theory suggested. They were awarded the 1978 Nobel Prize, getting all of the credit for the discovery of the cosmic microwave background. Their measurement combined with Hubble's discovery of the expansion of the universe now meant that the big bang model stood on solid ground.

The Cosmic Microwave Background

The cosmic microwave background (CMB) is thermal radiation that permeates all of the space in our universe, and is left over from a time when the universe cooled down enough for electrons to bond to nuclei. When the universe was hotter than 3000 K, all of the free electrons and nuclei resembled a dense fog and photons could not escape. As the universe expanded and cooled enough for electrons to bond to nuclei (forming hydrogen), it was as if the fog was cleared because photons interact weakly with hydrogen. This moment, about 380,000 years after the big bang, is called the "surface of last scattering" because it was the last time most of the CMB photons interacted with matter. The map[14] below shows seven years of observations of the cosmic microwave background from a satellite called the Wilkinson Microwave Anisotropy Probe (WMAP). This is an all-sky

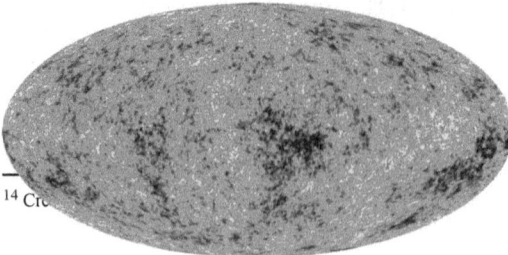

map: imagine taking this 2-dimensional image and wrapping it around your head, so that you can see different parts of the map in any direction that you look. This image shows a mapping of photons (or thermal radiation) that have been traveling in the universe uninterrupted for about 13.7 billion years. Remember that the temperature of the universe at the time of last scattering was about 3000 K, and that is equivalent to thermal radiation with a wavelength in the visible part of the spectrum. The expansion of the universe has caused the CMB photons to cool and their wavelengths to redshift by about 1000 times. Just as models predict, we observe a background radiation today that has a temperature of 2.73 K and a wavelength in the microwave part of the spectrum, it is a nearly perfect relic of the expected radiation from last scattering 380,000 years after the big bang.

The Cosmological Principle

The CMB observationally confirms the idea that the universe is isotropic and homogenous, a concept known as the cosmological principle. Isotropy means uniform properties in all directions, in other words, there are no preferred directions. Homogeneous means the same throughout, no matter where you observe from. Clearly, this is not true on small scales, for example your room looks different than your kitchen, landscapes in different areas of the Earth are different, and the part of the Galaxy that we live in looks different than the center of the Galaxy. However, the universe viewed on the largest scales (millions of light years) does appear isotropic and homogenous. The large scale structure of the universe to the east of us looks the same as the large scale structure to the west of us, and the density of the universe is the same everywhere. The cosmological principle also holds that the laws of physics are the same everywhere, meaning that the mass of an electron or the laws of gravity remain the same no matter where you are in the universe. A consequence of a homogenous and isotropic universe is that at one time it had to be causally connected, meaning that interactions were possible to allow conditions to equilibrate. This implies that the universe was much smaller and denser long ago, suggesting a creation. The best evidence we have for the cosmological principle is the cosmic microwave background observations appearing isotropic and homogenous, and the temperature of the background radiation appears the same in all parts of the CMB map.

Estimating the Age of the Universe

You might be wondering how astronomers know the age of the universe to be 13.7 billion years old. There are two main ways in which the age can be estimated: analyzing the oldest stars that we can observe or by using the expansion rate of the universe to extrapolate back to the big bang. Let's first discuss the oldest stars and stellar evolution.

The amount of light that a star gives off, its luminosity, is dependent on its mass; and luminosity and mass are related to the amount of time a given star has to burn through its fuel supply. For reference, our Sun can sustain for about 9 billion years before it runs out of fuel. A star that is twice the mass of the Sun will only live for about 800 million years, and a star that is 10 times the mass of the Sun will only have enough fuel to burn for 20 million years.

Luminosity is therefore directly related to mass and age of a star: a higher mass star will be brighter, bigger, and have a shorter lifetime than a lower mass star because it will burn through its fuel quicker. Astronomers look at globular clusters, an enormous collection of stars with high densities near the center, to put a lower limit on the age of the universe. Making the reasonable assumption that all of the stars in the cluster formed at about the same time, the age of the cluster can be estimated based on the stars that are still giving off light compared to the stars that are no longer shining. For example, if the cluster formed 2 billion years ago, all of the stars that have masses higher than twice the mass of the sun should no longer be burning their fuel. The oldest globular clusters are somewhere between 11-18 billion years old, which provides a lower limit to the age of the universe to be at least 11 billion years old. There is a significant amount of uncertainty in the estimate of the age of the globular clusters because we do not know the distance to the cluster precisely (which limits our ability to estimate its mass and luminosity) and there are uncertainties associated with stellar evolution models.

The second way that we can estimate the age of the universe is using the expansion rate and extrapolating back to the big bang. We already discussed Hubble's constant, which is a measurement of the expansion rate of the universe. It is likely that the expansion rate has changed over time, since the density and composition of the universe has changed as it evolved. Based on the range in the measured value of H_o, which is 65-80 km/s/Mpc, astronomers estimate a range in the age of the universe to be 12-14 billion years old. WMAPs observations of the cosmic microwave background has greatly enhanced our understanding of the density and composition of the universe, giving us a much better idea of the evolution of the expansion since the big bang. Using WMAPs measurements, we now have an incredibly accurate age estimate of 13.77 billion years ± 59 million years!

Abundance of Light Elements

It is now known that all of the elements in our universe were created in one of two ways:

1. Lighter elements (hydrogen, helium, lithium) were produced in the first few minutes after the big bang.
2. Elements heavier than helium were (and still are) produced inside stars and thus heavier elements were formed much later in the universe.

The big bang model predicts that the first chemical imprint in the universe was 75% hydrogen and 25% helium with trace amounts of other light elements, such as lithium. In the first few minutes after the big bang, a series of chemical reactions lead to production of twelve hydrogen atoms for every one helium atom. That is an atomic mass of twelve in hydrogen and four in helium. Thus, for each atomic mass of 16, we should expect the amount of hydrogen to be $\frac{12}{16} = \frac{3}{4} = 75\%$ of the total mass, and the amount of helium to be $\frac{4}{16} = \frac{1}{4} = 25\%$ of the total mass. This chemical process of converting protons, neutrons, and electrons into the light elements is called big bang nucleo-synthesis, because we are synthesizing atomic nuclei in the first moments after the big bang.

There is a lot of ongoing research focused on narrowing down the conditions in the universe that help us understand the chemical composition.

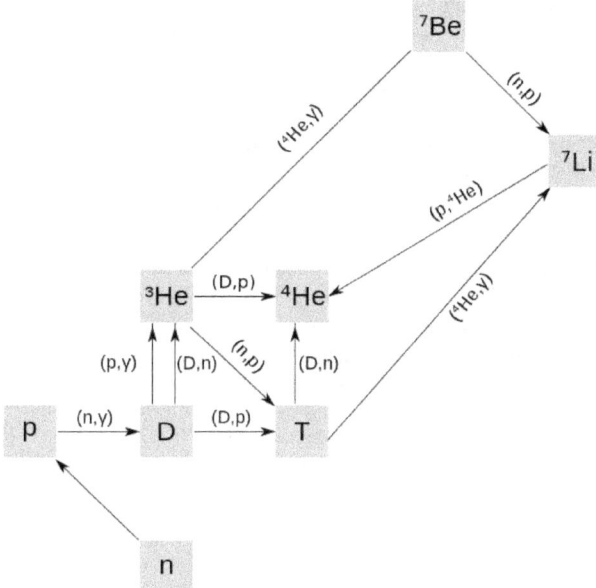

The nuclei of the light elements are hydrogen (one proton), deuterium (an isotope of hydrogen which has one proton and one neutron), helium-3 (two protons and one neutron), helium (two protons and two neutrons), and lithium (three protons and four neutrons). For example, a universe with a higher density of ordinary matter would end up with slightly more helium and lithium, but less helium-3 and deuterium. That is because of the simple fact that more chemical reactions could take place in a denser universe allowing more conversions of hydrogen into helium and lithium, while deuterium and helium-3 are carriers to allow that process to happen. Knowing the density allows us to narrow down the abundance of light elements to 75% hydrogen, 24% helium, and trace amounts of lithium.

Elements heavier than helium are made in stars. For example, the carbon and oxygen in our bodies were not produced in the universe until the first stars synthesized them, about one billion years after the big bang. The big bang gave us a chemical starting point, and then stars took over enriching the universe with heavier elements through fusion. Stars with masses close to our Sun will burn hydrogen and helium and leave behind an inert carbon core. A strong stellar wind will eject gases from the star, a stage of evolution called a planetary nebula, enriching the space around it. Stars that form from gas enriched by planetary nebulae will have higher abundance of heavy elements. Massive stars synthesize helium, carbon, oxygen, silicon, sulfur, and leave behind an inert iron core. If a star is massive enough, it will end its life with a supernova, which is a giant catastrophic explosion ejecting most of the mass of the star into space, and in process producing elements heavier than iron. It's amazing to consider that cosmic catastrophes are required to fuse most of the elements we find on Earth and in the universe!

This embodies two fields of astronomy that scientists spend their careers on: stellar nucleosynthesis and stellar evolution. Years of research in these fields provide an impressive description of how stars and supernovae contribute to the vast cosmic recycling of the elements in the universe necessary to form planets and living things.

Inflation Theory

Although the big bang model has greatly advanced our understanding of cosmology, there are still a few unexplained phenomena, namely:
1. Where does the large-scale structure come from?
2. Why is the overall distribution of matter so uniform?
3. Why is the density of the universe so close to the critical density?

An early episode of rapid inflation can solve all of these mysteries, and it is now widely accepted that inflation is an integral piece of the big bang model. Inflation occurred in the first tiny fractions of a second after the big bang, when the universe expanded exponentially by a factor[15] of about 10^{26}.

Large Scale Structure

The large-scale structure of the universe is a map of all of the visible or non-visible sources in the sky, such as galaxies, voids, galaxy superclusters, and filaments. It can be interpreted as

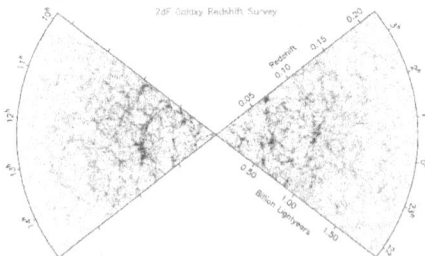

density fluctuations on the largest scales in the universe, and it has been surveyed and modeled by different teams of scientists. For example, one team carried out the 2dF galaxy redshift survey[16], lead by the Anglo-Australian Observatory from 1997-2002. They observed two large areas of the sky out to a distance of about 2.5 billion light years. The image to the left shows the large scale structure from now (the center of the image, where the slices intersect) to 2.5 billion years ago (remember the

---

[15] The factor that the universe grew during the inflationary period is not known, some sources estimate factors much higher than stated here. Value quoted here is an estimate from the NASA/WMAP team.
[16] More information on the 2dF Survey can be found at http://www.2dfgrs.net

universe is 13.7 billion years old!). Notice the blue areas where matter (galaxies, superclusters, stars, gas, etc.) piles up, and white areas where space appears to be empty (voids). Inflation theory explains that the quantum ripples in the tiny universe before exponential expansion were the seeds of the structure observed in the universe today.

The Horizon Problem

In any direction we look, the temperature and density of the cosmic microwave background radiation is the same, even though distances between opposite edges of the universe are too far apart for them to communicate. This is called the horizon problem. Consider a galaxy to the "west" that is 10 billion light years away and a different galaxy to the "east" that is also 10 billion light years. Since the universe is 13.7 billion years old, photons from each galaxy have had time to travel to us but have not had nearly enough time to travel the 20 billion light years to each other. That means that the east and west galaxies have not had any communication or interaction, and one might expect that the properties would be different. If, for example, the temperatures in the universe varied even a little bit, we should still see that variation since galaxy west and galaxy east could not interact to reach an equilibrium temperature. What we observe is the opposite: the universe is isotropic and homogeneous, meaning that the temperature of the universe is 2.728 ± 0.004 K anywhere we look. That is a very accurate measurement of the temperature from the microwave background from WMAP. Inflation theory attempts to solve this problem by assuming that the properties in the early universe equilibrated before rapid expansion when the universe was small, causing the uniform properties to be set at the time of inflation. That means that the building blocks from which galaxy west and galaxy east formed from were part of the same cosmic soup, interacting and equilibrating early-on, and then evolving from the same initial conditions without any interaction with each other after the release of the microwave background, making it plausible that they would end up in a very similar state.

The Flatness Problem

The initial conditions of the big bang model have huge effects on the properties and future of the universe in such a way that the universe appears to be fine-tuned. One of the parameters is the density of matter and energy, which affects the overall geometry of the universe. If the density of the universe was just 10% higher, it would have collapsed a long time ago (closed geometry); and if the density was just 10% lower, galaxies would never have formed (open geometry). The critical density is the sweet spot, defining what we call a flat universe with enough density to allow matter to interact gravitationally but not so much that it causes the universe to re-collapse on itself. The density of the universe is so close to this critical density that it appears to be an unbelievable coincidence, this is the flatness problem. Inflation theorists try to explain this problem by arguing that no matter the initial geometry of the universe, the rapid expansion would eventually cause space to flatten out, forcing the density towards the critical density.

Fate of the Universe

The major forces involved in the fate of the universe are the momentum of expansion and the pull of gravity. The image[17] below shows one possible scenarios for the evolution of the universe, with time on the vertical axis. If gravity wins, we experience "The Big Crunch", the universe re-collapsing on itself; but if expansion wins, we may find ourselves in "The Big Freeze", where the universe expands and cools until it cannot sustain star, planet, or life formation, which will happen below. The Big Freeze also results in what is known as the "Heat death" of the Universe. At some point, there is a uniform temperature of all matter. Since there is no temperature differential, no thermodynamic "work" can be done.

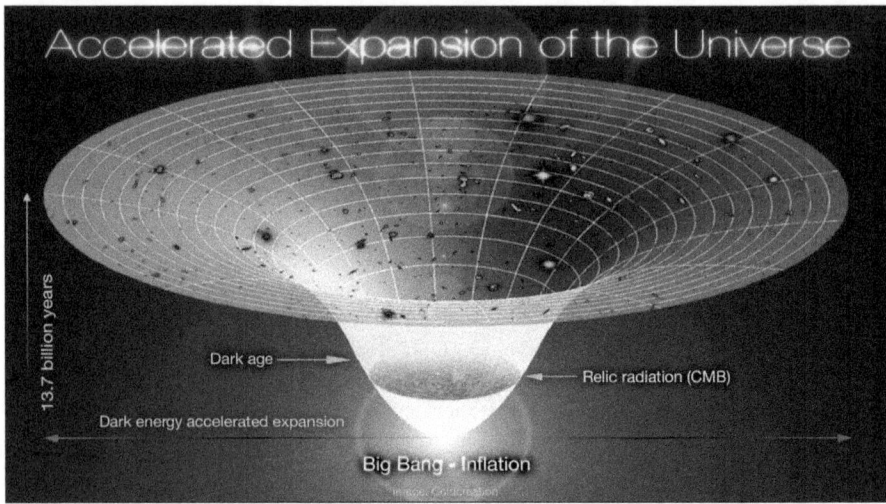

A universe with a density equal to the critical density will not collapse, but will expand more and more slowly over time (this is the universe with flat geometry discussed in the previous section, and it is shown in the image below as the "critical universe"). A coasting universe has a density higher than critical and will expand forever without slowing down. The accelerating universe is one that expands faster and faster with time, which is not easily explained with just gravity and the momentum of expansion. Observations of distant supernova explosions suggest that the universe is likely accelerating, meaning that the expansion of space appears to be speeding up with time. If the accelerating universe model is correct, then the momentum of expansion of the universe is not enough to explain our observations, and we have to introduce an additional repulsive energy, called dark energy, causing the expansion of the universe to speed up. While the fate of the universe is still unknown and astronomers know almost nothing about

---

[17] Image Credit: NASA

dark energy, but observations allow us rule out re-collapse as an option. The way in which our universe will expand is a subject of ongoing research.

Alternate Theories

The Big Bang is the most popular theory, for good reason, but it is still just a theory. If some of our basic assumptions are incorrect, it is possible, but unlikely, that the big bang model is incorrect and one of the other competing models describe the physical universe better. The other possible models are (1) the steady-state universe, (2) eternal inflation, or (3) an oscillating universe.

In the 1940's, astronomer Fred Hoyle developed an alternate theory to the big bang theory called the steady-state universe. The premise for this new model is that the density of matter in the universe does not change because matter is created as the universe expands. The theory states that the universe obeys a perfect cosmological principle, meaning that the properties are the same at any place or at any time. This theory had a lot of attention in the 1940's, but has been rejected as a result of the following observational tests. Today, we know that the universe has changed, and one such example of that is the observation of quasars (quasi-stellar radio source). Quasars are all very distant objects, and are extremely bright radio sources emitting large amounts of energy. They are assumed to be emission from massive black holes in the center of galaxies, and may be a stage in galaxy evolution. The fact that we do not observe quasars in the local universe is evidence against the steady-state model. "The final nail in the coffin", as Stephen Hawking said, for the steady-state model was the observation of the cosmic microwave background. The assumption in the steady-state model was that the cosmic microwave background was produced from light from stars that has been scattered by dust in galaxies. However, the microwave radiation is too uniform in all directions for it to be scattered star light, and additionally it shows no polarization that we should see if light is scattered off of dust. The steady-state model has been largely rejected by astronomers and has given the big bang model steam to be tested and modified for the better.

The eternal inflation model is an extension of the big bang model and the premise is that some parts of the universe are expanding exponentially forever. The major implication of the eternal inflation model is the existence of the multiverse, the hypothesis that there are many (possibly infinite) universes, other than our own. There is no evidence of this idea and there are several different theories about the properties of potential other universes, and the inflationary universe model itself.

The oscillating universe combines two theories for the evolution and fate of the universe: the big bang theory and the big crunch theory. The assumption is that the universe will oscillate between the big bang and the big crunch, continually expanding and contracting forever. This theory was proposed by Richard Tolman in the 1920's. The idea is that the expansion of the universe will eventually slow down to a halt, begin contracting until it crunches in on itself. This cycle begins all over again with a big bang, and so on. The major evidence against the oscillating universe theory is that recent supernova data has confirmed a model for the fate of the universe that will not contract back on itself. In fact, we now have evidence that the universe will expand forever, as discussed in the previous section.

Dark Matter and Dark Energy

The first evidence of dark matter was discovered in the 1930's, by Swiss astronomer Fritz Zwicky. At that time, the field of astronomy was not developed enough to immediately understand dark matter, so it took another forty years for the "dark matter problem" to be an active field of research. To put it in perspective, the Great Debate was only a decade earlier, when it was unclear whether or not the observed "spiral nebulae" were other galaxies or if the Milky Way was the whole universe. Einstein's theory of general relativity was just beginning to be tested, and the big bang theory was in the earliest stages of development.

Zwicky was working at the California Institute of Technology (Cal Tech) observing the gravitational effects of the galaxies within the Coma Cluster. The Coma Cluster is a large collection of galaxies (~1000) that are gravitationally bound to each other. Zwicky convinced Cal Tech to build him a new telescope specifically for the purpose of his project: to be able to capture the entire Coma Cluster in the field of view and calculate the radial velocities from the measurement of the redshift in the spectrum of each galaxy. Zwicky used the virial theorem, a well-known theory of classical mechanics, to relate the velocity of the galaxies to their gravitational force. From the gravitational force, Zwicky calculated the total mass of the Coma Cluster. He also measured how much total light is given off by the cluster as a whole, which is made up of trillions of stars in the thousands of galaxies. Zwicky found that the total light output per unit mass is far below what was expected by over a factor of 100, leading him to conclude that there was extra unseen mass in the Coma Cluster and he called it "dark matter".

About 40 years later, in 1973, two astronomers named Jeremiah Ostriker and James Peebles simulated the evolution of galaxies by programming the dynamics of particles. In this type of theoretical work, called N-body simulation, the two astronomers defined many mass particles (where each mass particle represents a collection of stars, dust, and gas) rotating about a central point. Their simulation calculated the gravitational force between each pair of mass particles and determined the next location for each particle. The simulation would repeat this exercise in small time steps over a long period of time. Theoretical models are tested by comparing the results of the simulation to real observations. Ostriker and Peebles were able to reproduce observations of galaxies only if they added a static, uniform distribution of mass several times the size of the total mass of the visible galaxy. This was the first numerical evidence for dark matter.

Another key to the puzzle came in at about the same time when astronomers Kent Ford and Vera Rubin at the Carnegie Institute of Washington did an observational study of the Andromeda Galaxy (also called Messier 31 or M31), our nearest major galactic neighbor. They measured the motion of hydrogen gas clouds that orbit just like the stars in the galaxy. The expectation was that the further out the gas was from the center of the galaxy, the slower it would move in its orbit. This is the prediction of the virial theorem if the mass in the galaxy were concentrated at the center. Light given off from the galaxy is concentrated at the center, so if we expect that the light is emitted from mass, the mass should be concentrated at the center as well. Their observations showed the opposite, the orbital velocity of the gas clouds did not go down with distance from the center, but instead appeared to remain constant. Their conclusion was that if Newton's Laws were correct, then there must be dark matter mass that increased with distance from the center of the galaxy. The figure[18] to the left shows the results from different observations of the Triangulum galaxy (also called M33). The plot shows rotational velocity as a function of the distance, R, from the center of the galaxy. This type of plot is called a rotation curve. The yellow dots with error bars are the data from the stars and the blue dots with error bars are data from the 21-cm spectral line of hydrogen gas. The dashed line shows the expected rotation from the visible disk, assuming that all of the mass in the galaxy is concentrated where the light concentrated. The observations show that the data do not fit the model that includes mass from a visible disk alone. Instead, the data fits well with a model that include mass from both a visible disk and a dark matter halo, shown as the solid line. The dark matter halo that fits the data has a mass distribution that increases with distance from the center.

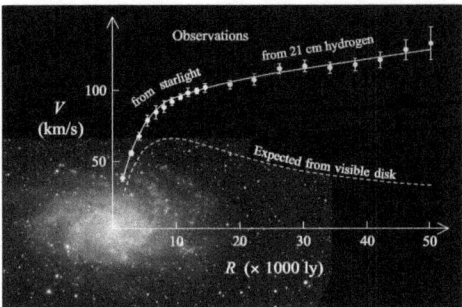

By the end of the 1970's, there was clear evidence for dark matter in both observations of hydrogen gas clouds orbiting in galaxies and in observations of the gravitational effects and mass-to-light ratios of large clusters of galaxies (such as the Coma Cluster discussed earlier). Further studies of simulated galaxy evolution showed that significantly more mass than the luminous mass was needed in order to make spiral and elliptical galaxies evolve to match observations. The two possible explanations for the discrepancy in the amount of luminous mass was either that our fundamental understanding of gravity was wrong or there was much more unseen mass in the universe than we thought. There were significant attempts to redefine gravity and modify Newton's laws.

More evidence developed in the 1990's from observations of the cosmic microwave

---

[18] Data from E. Corbelli, P. Salucci (2000). "The Extended Rotation Curve and the Dark Matter Halo of M33". Monthly Notices of the Royal Astronomical Society, 311, 441-447. https://arxiv.org/abs/astro-ph/9909252

background (CMB). Recall from the map of the CMB on page 13 that there are very small temperature variations that are represented as different colors on the map. The map as a whole is an observation of the last scattering event of photons 380,000 years after the big bang when the temperature of the universe was around 3000 K. Before the universe cooled to 3000 K, it was too hot for electrons to remain bound to nuclei, so there were no stable atoms and the universe was opaque because photons continually scattered off of free electrons and nuclei. During this epoch, before neutral hydrogen formed, matter was distributed almost uniformly with very small fluctuations. The constant scattering left matter in a state of thermal equilibrium, and close to a uniform temperature. The quantum mechanical fluctuations created small density fluctuations that pulled matter (both normal and dark matter) towards the center of the fluctuations where the density was slightly higher. Both normal matter and dark matter are affected by gravity the same way, but dark matter does not interact with photons while normal matter does. Therefore, as both types of matter are pulled toward higher density, normal matter is pushed back out by photon pressure while dark matter sinks toward the center of the fluctuation, increasing the total mass in that volume. Normal matter is pushed out until gravity overcomes photon pressure, and then it sinks back in. This created a ringing effect, and the normal matter heats up as it is pulled in and cools down when it is further from the higher density region. Dark matter is unaffected by this process. Once hydrogen atoms could form, areas that were hotter from this ringing process tracked areas of higher density and higher concentrations of dark matter.

The temperature fluctuations from the ringing were set at the time of last scattering when neutral hydrogen formed and the universe became transparent. The variations in temperature from this process can be observed and studied in the map of the cosmic microwave background (CMB). The image[19] to the left shows the power spectrum of the CMB on different angular

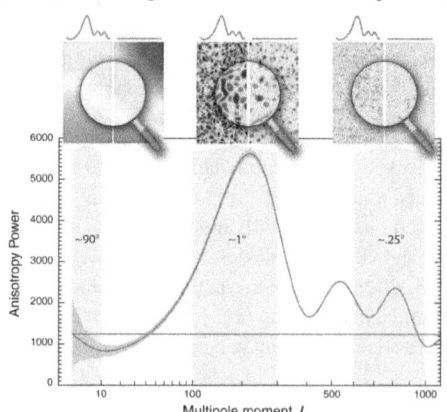

scales. A power spectrum is the amount of power or energy per unit time as a function of frequency or angular size scale. The red curve represents data and the straight red line shows the power spectrum with a hypothetical constant temperature. The images above the plot are showing the sky maps at each of the highlighted angular scales. Each little image shows how the map should look if the temperature varies (left image) versus a constant temperature (right image). The map above the angular scale of 90° shows variation in power (or temperature) on large scales, while the

---

[19] Power spectrum of the cosmic microwave background from NASA/WMAP team. More information can be found at http://map.gsfc.nasa.gov/mission/sgoals_parameters_spect.html

highlighted area in the middle shows the variation in power with angular size scales of 1°, and the variation in power on the smallest scales is shown in the right panel. The overall fit to the observations that make up the red curve are dependent on the amount of ordinary matter, the amount of dark matter, the amount of dark energy, the Hubble constant, the redshift (or distance) of the surface of last scattering, and the strength of the initial ripples in temperature. So by fitting the model with these parameters to the data, astronomers estimate the amount of dark matter in the universe to be 23%, the amount of dark energy to be 73%, and the amount of ordinary matter to be 4%.

The real smoking gun came in 2006 when a group of astronomers observed the aftermath of the collision of two galaxy clusters. Since dark matter does not interact much with either itself or with ordinary matter, it remains unaffected after a collision. Ordinary matter, on the other hand, does interact with itself and dark matter. So when the two clusters collided, the dark matter passed through unaffected while the ordinary matter is pulled forward by the dark matter in its own cluster and also pulled back by the dark matter plus the ordinary matter in the other cluster. The net effect is that the ordinary matter lags behind the dark matter of its own cluster. The image[20] to the left shows an x-ray and visible light composite image of the Bullet Cluster. During the collision, the normal matter heats up and produces x-ray emission, which is shown in red. Gravitational lensing experiments track where most of the mass is in the cluster, since large fields of gravity distort the image of the background galaxies. The concentration of the mass in the cluster is shown in blue, which is clearly separated from the ordinary matter (red), providing evidence that most of the matter in the cluster is dark matter. This experiment gave astronomers confidence in their understanding of gravity and that dark matter truly exists.

There is not as much known about dark energy as there is about dark matter. The existence of dark energy was proposed in the 1990's when the expansion of the universe appeared to be accelerating, and it could not be explained without another form of energy. Cosmological models discussed earlier, namely the fit to the power spectrum of the cosmic microwave background indicates that dark energy makes up the majority of total energy in the universe. The best hypotheses for dark energy that we have are that either

---

[20] Evidence for dark matter in the Bullet Cluster is shown in the x-ray and visible light composite. Credit: X-ray: NASA/CXC/CfA/M.Markevitch et al.; Optical: NASA/STScI; Magellan/U.Arizona/D.Clowe et al.; Lensing Map: NASA/STScI; ESO WFI; Magellan/U.Arizona/D.Clowe et al.

(1) Dark energy may be another fundamental force of nature, like gravity, that only comes into effect when the universe reaches a certain size or perhaps a temporary force.

(2) Dark energy connects the long-standing question of how to connect the physics of the very small (quantum mechanics) to the very large (Einstein's theory of gravity).

Our understanding of dark energy is in its infancy, but we do know that it is a force that permeates space everywhere at a very low density, and it appears to have a large effect on the expansion of the universe. The unknown dark energy force signifies that we still have a lot to learn about physics in the universe.

Future Experiments

It is truly incredible that we know anything about the cosmos, and even more incredible that we can measure faint signals and progress our understanding of the universe with some certainty. Astronomy is a very active field of ongoing research and has a lot of exciting projects on the horizon. The National Aeronautics and Space Administration (NASA) is just one of many large space agencies dedicated to understanding the evolution of the universe, others include the European Space Agency (ESA), the National Optical Astronomy Observatory (NOAO), and the Jet Propulsion Lab (JPL), to name just a few. There are many more government agencies in the world and throughout universities that share similar goals.

NASA has a program dedicated to understanding our origins in their Cosmic Origins Program. Some of their science goals include understanding the properties of the first stars in the universe, how dark matter evolved and influenced galaxy formation, how galaxy evolution led to our own galaxy, when supermassive black holes formed and how they affect galaxy evolution, and how stars and planetary systems formed and evolved. Another program on the horizon is the Wide Field Infrared Survey Telescope, or WFIRST, a space telescope designed to survey the sky in the infrared. WFIRST is a 2.4 meter telescope (about the same size as the Hubble Space Telescope) planned for launch in 2020. One of its science objectives is to understand dark energy and the acceleration of the expansion rate of the universe. The James Webb Space Telescope (JWST) is another infrared satellite planned for launch in October 2018. The overarching goal of JWST is understanding our origins better. Its science goals are understanding the epoch of reionization, the first stars and galaxies, galaxy evolution, the birth of stars and protoplanetary systems, and the origins of life.

Conclusion

It is hard to wrap our minds around the fact that there are 100 billion galaxies scattered about in a universe 46 billion light years across. There are about 100,000 billion billion stars; and our galaxy, the Milky Way, has about 400 billion stars. The building blocks of our universe are hydrogen and helium, and there are a billion photons for every particle. We live on an average planet around an average star, in an average galaxy, in a very large universe. It is mind-boggling to imagine some or any of these conditions being different and how that would impact the current state of the universe and life as we know it. The universe is an amazing place, and we are incredibly lucky to be alive in it, let alone have what it takes to engineer technology to observe it at its greatest reaches.

Online Resources

Other books about space by Charles River Editors

Other books about the Universe on Amazon

Bibliography

Barrow, J. D. (1994). The Origin of the Universe. Weidenfeld & Nicolson. ISBN 0-297-81497-4.

Bartel, Leendert van der Waerden (1987). "The Heliocentric System in Greek, Persian and Hindu Astronomy". Annals of the New York Academy of Sciences. 500 (1): 525–545. Bibcode:1987NYASA.500..525V. doi:10.1111/j.1749-6632.1987.tb37224.x.

"Cosmic Journey: A History of Scientific Cosmology". American Institute of Physics.

Davies, P. C. W. (1992). The Mind of God: The Scientific Basis for a Rational World. Simon & Schuster. ISBN 0-671-71069-9.

Feuerbacher, B.; Scranton, R. (2006). "Evidence for the Big Bang". TalkOrigins.

Landau L, Lifshitz E (1975). The Classical Theory of Fields (Course of Theoretical Physics). 2 (revised 4th English ed.). New York: Pergamon Press. pp. 358–397. ISBN 978-0-08-018176-9.

Liddell, H. G. & Scott, R. (1968). A Greek-English Lexicon. Oxford University Press. ISBN 0-19-864214-8.

Mather, J. C.; Boslough, J. (1996). The Very First Light: The True Inside Story of the Scientific Journey Back to the Dawn of the Universe. Basic Books. p. 300. ISBN 0-465-01575-1.

Misner, C.W., Thorne, Kip, Wheeler, J.A. (1973). Gravitation. San Francisco: W. H. Freeman. pp. 703–816. ISBN 978-0-7167-0344-0.

Raine, D. J.; Thomas, E. G. (2001). An Introduction to the Science of Cosmology. Institute of Physics Publishing.

Rindler, W. (1977). Essential Relativity: Special, General, and Cosmological. New York: Springer Verlag. pp. 193–244. ISBN 0-387-10090-3.Alpher, R. A.; Herman, R. (1988). "Reflections on Early Work on 'Big Bang' Cosmology". Physics Today. 8 (8): 24–34. Bibcode:1988PhT....41h..24A. doi:10.1063/1.881126.

Riordan, Michael; William A. Zajc (2006). "The First Few Microseconds" (PDF). Scientific American. Nature Publishing Group. 294 (5): 34–41. doi:10.1038/scientificamerican0506-34a. Archived from the original (PDF) on 30 November 2014.

Singh, S. (2004). Big Bang: The Origins of the Universe. Fourth Estate. ISBN 0-00-716220-0.

"Misconceptions about the Big Bang" (PDF). Scientific American. March 2005.

Weinberg, S. (1993). The First Three Minutes: A Modern View of the Origin of the Universe. Basic Books. ISBN 0-465-02437-8.

Free Books by Charles River Editors

We have brand new titles available for free most days of the week. To see which of our titles are currently free, click on this link.

Discounted Books by Charles River Editors

We have titles at a discount price of just 99 cents everyday. To see which of our titles are currently 99 cents, click on this link.

www.ingramcontent.com/pod-product-compliance
Lightning Source LLC
Chambersburg PA
CBHW061235180526
45170CB00003B/1305